The Solar System

by Margaret J. Goldstein

⌐ Lerner Publications Company • Minneapolis

Text copyright © 2003 by Margaret J. Goldstein
Reprinted in 2006

Lerner Publications Company
A division of Lerner Publishing Group
241 First Avenue North
Minneapolis, MN 55401 USA

Website address: www.lernerbooks.com

Words in **bold type** are explained in a glossary on page 30.

Library of Congress Cataloging-in-Publication Data

Goldstein, Margaret J.
    The solar system / by Margaret J. Goldstein.
        p.   cm. – (Our universe)
    Includes index.
    Summary: An introduction to the bodies in the solar system, including the Sun, planets, asteroids, meteoroids, and comets.
        ISBN-13: 978–0–8225–4657–3 (lib. bdg. : alk. paper)
        ISBN-10: 0–8225–4657–4 (lib. bdg. : alk. paper)
    1. Solar system–Juvenile literature. [1. Solar system.]
    I. Title. II. Series.
    QB501.3 .G65 2003
    523.2–dc21   2002004712

Manufactured in the United States of America
2 3 4 5 6 7 – JR – 11 10 09 08 07 06

The photographs in this book are reproduced with permission from: © Roger Ressmeyer/COR-BIS, p. 3; NASA pp. 4, 10, 11, 12, 13, 14, 15, 17, 18, 19, 20, 21, 22, 25; © ESA/Tsado/Tom Stack & Associates, p. 5; © USGS/Tsado/Tom Stack & Associates, p. 16; © JPL/Tsado/Tom Stack & Associates, p. 23; © Tsado/NASA/Tom Stack & Associates, p. 24 © John Sanford/Photo Network, p. 27.

Cover: NASA.

The Sun shines in the daytime. At night
the Moon glows. Planets twinkle
among the stars. Where are the Sun,
the Moon, and the planets?

The Sun, Moon, and planets are in the solar system. The solar system stretches across space for billions of miles. Earth is part of the solar system. Earth is our home planet.

The Sun is at the center of the solar system. The Sun is a glowing star. It shines hot and bright. It is much bigger than anything else in the solar system.

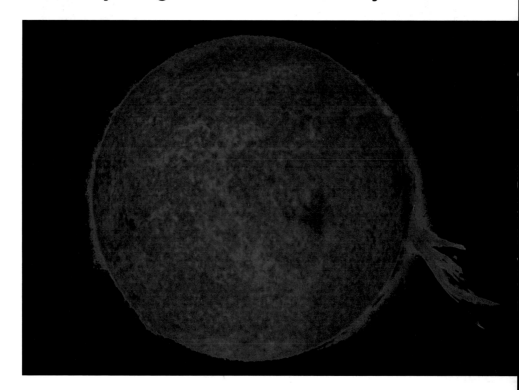

Nine planets travel around the Sun.
Their names are Mercury, Venus, Earth,
Mars, Jupiter, Saturn, Uranus, Neptune,
and Pluto.

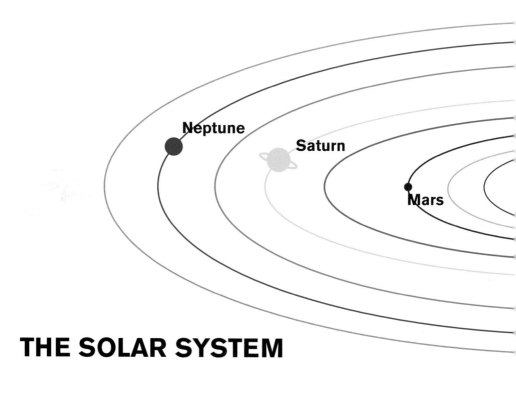

THE SOLAR SYSTEM

Each planet takes a different path around the Sun. The paths are called **orbits.** Most planets have an oval orbit.

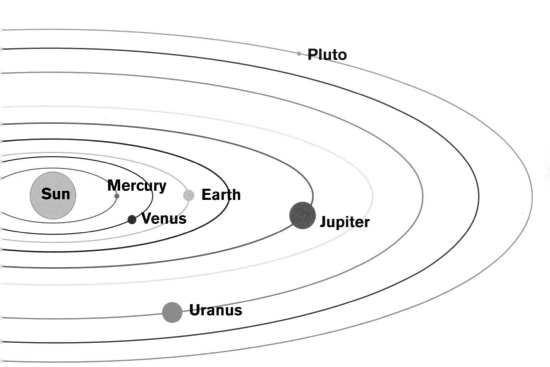

The planets also spin around like tops. This spinning is called **rotating.**

Each planet rotates around an imaginary line called an **axis.** The axis goes through the center of the planet. Most planets have a tilted axis.

What are the planets made of?

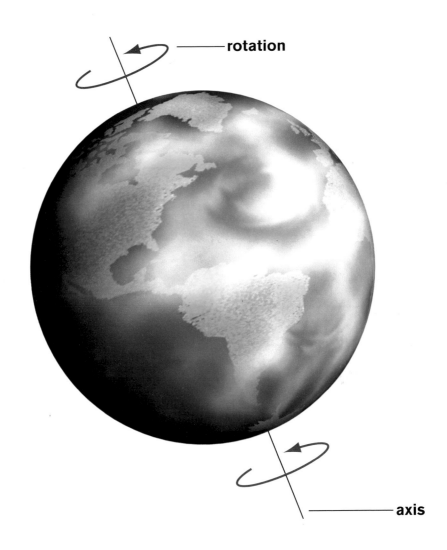

rotation

axis

Some of the planets in the solar system are made of rock and metal. These rocky planets are Mercury, Venus, Earth, Mars, and Pluto. Their ground is solid and hard.

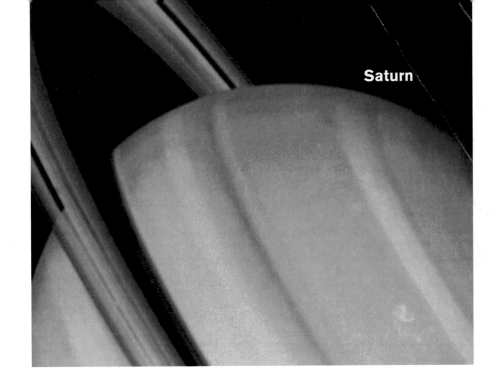

Saturn

Jupiter, Saturn, Uranus, and Neptune
are made mostly of gases. These
planets are much bigger than the
others. Each gas planet is circled by
rings made of ice, dust, and rock.

Mercury is the closest planet to the Sun. It bakes in the Sun's heat during the day. Its ground is covered with wide holes called **craters.** Mercury has no wind, rain, or clouds.

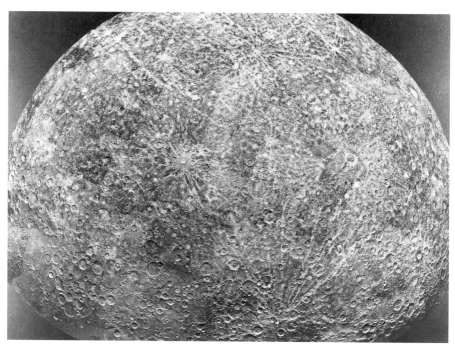

The second planet is Venus. Venus is surrounded by a thick layer of gases. This layer is called an **atmosphere.** Venus's atmosphere traps the Sun's heat. So Venus is always burning hot.

Earth is third from the Sun. Most of our planet is covered with water. Plants, animals, and people all live on Earth. Earth is the only planet in the solar system known to have living things.

A smaller body circles around Earth.
Do you know what it is? It is the Moon.
Some planets have no moons. Other
planets have many moons.

Mars is the fourth planet from the Sun. It is nicknamed the Red Planet. Red rocks cover the ground on Mars. And red dust swirls in the wind. Two moons circle Mars.

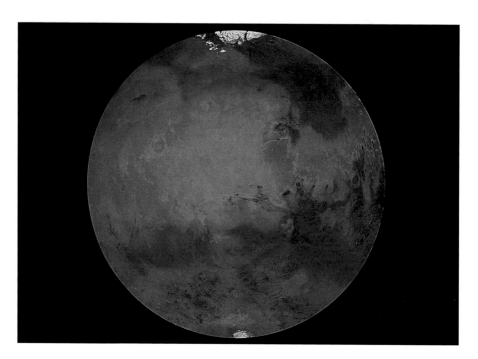

The fifth planet is Jupiter. It is the largest planet by far. Its atmosphere is stormy and windy. The biggest storm is called the Great Red Spot. Jupiter has 4 rings and at least 39 moons.

Saturn is sixth from the Sun. Wide, flat rings circle this planet. Saturn's rings are the largest and brightest rings in the solar system. Saturn also has at least 30 moons.

The seventh planet is Uranus. It is a pale blue-green planet. Circling Uranus are 11 rings. They are not bright enough to be seen clearly. At least 21 moons also circle Uranus.

Neptune is usually the eighth planet from the Sun. It is a big blue ball of gas. Strong winds and bright clouds whip through Neptune's atmosphere.
Neptune has four rings and eight moons.

Pluto is usually the farthest planet from the Sun. It is a cold, icy ball of rock. It is the smallest planet in the solar system. Pluto has one moon called Charon.

Many other objects circle the Sun.
Comets are small bodies made of rock, ice, and dust. They often have long glowing tails. They shine brightly when they come close to the Sun.

Asteroids travel around the Sun, too. They are like tiny, rocky planets. But most of them are not round. Many asteroids travel in the asteroid belt between Mars and Jupiter.

Small chunks of rock and metal also circle the Sun. The chunks are called **meteoroids.** Sometimes meteoroids, comets, and asteroids crash into planets or moons and form craters.

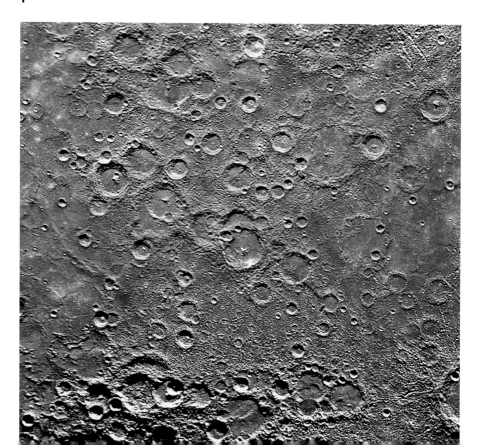

People have studied the solar system using telescopes, cameras, and other machines. Spacecraft have visited most of the planets. **Astronauts** have visited the Moon.

People have also studied space beyond the solar system. There are billions and billions of stars in outer space. There might be billions of other solar systems, too.

Imagine taking a tour of the solar system. Which planets and moons would you choose to visit?

Facts about the Solar System

Planet	Distance from Sun	Diameter (distance across)
Mercury	36,000,000 miles (58,000,000 km)	3,030 miles (4,880 km)
Venus	67,200,000 miles (108,000,000 km)	7,520 miles (12,100 km)
Earth	93,000,000 miles (150,000,000 km)	7,930 miles (12,700 km)
Mars	142,000,000 miles (228,000,000 km)	4,220 miles (6,790 km)
Jupiter	484,000,000 miles (778,000,000 km)	88,7000 miles (143,000 km)
Saturn	887,000,000 miles (1,430,000,000 km)	74,900 miles (121,000 km)
Uranus	1,780,000,000 miles (2,880,000,000 km)	31,800 miles (51,100 km)
Neptune	2,800,000,000 miles (4,500,000,000 km)	30,800 miles (49,500 km)
Pluto	3,670,000,000 miles (5,900,000,000 km)	1,430 miles (2,300 km)

Orbit Period	Rotation Period
88 days	60 days
225 days	243 days
365 days	24 hours
687 days	25 hours
12 years	10 hours
29 years	11 hours
84 years	17 hours
165 years	16 hours
284 years	6 days

Glossary

asteroids: small chunks of rock or metal that circle the Sun

astronaut: a person who travels into space

atmosphere: the layer of gases that surrounds a planet or moon

axis: an imaginary line that goes through the center of a planet

comets: small bodies made of dust, gas, and ice. A comet has a shining tail when it comes close to the Sun.

craters: large holes on a planet or moon

meteoroids: small pieces of rock or metal that circle the Sun

orbit: the path of a small body that travels around a larger body in space

rotating: spinning around in space

Learn More about the Solar System

Books

Rau, Dana Meachen. *The Solar System.* Minneapolis: Compass Point Books, 2001.

Simon, Seymour. *Our Solar System.* New York: Morrow, 1992.

Theodorou, Rod. *Across the Solar System.* Chicago: Reed Educational & Professional Publishing, 2000.

Websites

Solar System Exploration
http://solarsystem.nasa.gov/index.cfm
Detailed information from the National Aeronautics and Space Administration (NASA) about all the bodies in the solar system, with good links to other helpful websites.

The Space Place
http://spaceplace.jpl.nasa.gov
An astronomy website for kids developed by NASA's Jet Propulsion Laboratory.

StarChild
http://starchild.gsfc.nasa.gov/docs/StarChild/StarChild.html
An online learning center for young astronomers, sponsored by NASA.

Index